Mibson Michel Santiago Ramos
Felipe Jose S. A. Melo
Marcelo T. Gurgel

Geology Applied to Civil Engineering in the Municipality of Mossoró - RN

Mibson Michel Santiago Ramos
Felipe Jose S. A. Melo
Marcelo T. Gurgel

Geology Applied to Civil Engineering in the Municipality of Mossoró - RN

Main Methods or Techniques of Geological Studies in Civil Construction

ScienciaScripts

Imprint
Any brand names and product names mentioned in this book are subject to trademark, brand or patent protection and are trademarks or registered trademarks of their respective holders. The use of brand names, product names, common names, trade names, product descriptions etc. even without a particular marking in this work is in no way to be construed to mean that such names may be regarded as unrestricted in respect of trademark and brand protection legislation and could thus be used by anyone.

Cover image: www.ingimage.com

This book is a translation from the original published under ISBN 978-613-9-69185-2.

Publisher:
Sciencia Scripts
is a trademark of
Dodo Books Indian Ocean Ltd. and OmniScriptum S.R.L publishing group

120 High Road, East Finchley, London, N2 9ED, United Kingdom
Str. Armeneasca 28/1, office 1, Chisinau MD-2012, Republic of Moldova, Europe
Printed at: see last page
ISBN: 978-620-8-12579-0

DEDICATORY

To my parents, for all their support, patience, understanding and dedication during my course, a period in which they placed all their trust in me that everything would turn out as desired, serving as a great learning experience for me personally.

ACKNOWLEDGEMENTS

To my mother, Osineide Santiago, for all her love, encouragement and trust; for all the advice and teachings that have made me the person I am today; for often giving up her wishes to make mine possible; my eternal thanks for her tireless dedication to me; for being my safe haven and my greatest example of character and willpower.

To my father, Waldemar Ramos, for always being so present in my life, allowing me to lack nothing. Thank you for your trust in me, for your support in difficult times and for being so tough and generous with me when I needed it.

To my sister, Misma Michelle, for her company and moments of relaxation; for sharing so many happy occasions with me, which allowed us to overcome our differences.

To my brother-in-law, Jailson Felix, for being almost like a brother, helping me out of everyday difficulties that were considered difficult for me.

To my girlfriend, Marcelle Nunes, who has often helped me and wiped away my tears in the face of the adversities I've faced in my academic and social life, supporting me and helping me to carry on with her love and being the key to this Monograph being completed.

To my grandmother, Maria Cordeiro, who has always been present in my life with her advice and prayers, supporting me and rooting for my success. You are an example of kindness and wisdom; thank you for encouraging me with your wise and comforting words, I can only thank you for your unconditional affection grandma.

To all my family, who are my foundation; my reference point for unity, love, trust and support.

To my university friends, for the true friendship we've built over the years; with you I've shared everything from the moments of distress before exams to the moments of immense happiness at each stage completed; I thank you for your patience, for the late nights studying and supporting each other, and for the words of encouragement that didn't allow me to give up.

To the undergraduate teachers who were part of this journey, contributing to the completion of this work and consequently my professional training. In particular, Jacqueline Cunha de Vasconcelos Martins, who guided me on my first monograph, Francisco Praxedes de Aquino and Marcelo Tavares Gurgel, who proved to be true friends more than teachers, and Mayara Raissa Medeiros Rodrigues, a building technician at the Federal Institute of Science and Technology of Rio Grande do Norte (IFRN), who provided important information and material on the geology of Mossoró.

To the high school teachers - in the name of José Valter Rebouças and Lúcia de Fátima Tôrres Câmara Alves - people who are the basis for my having come this far. I will always be grateful for the words that often opened my eyes and contributed to my growth. You believed in my potential and that's why I was able to believe in it too; thank you for your significant contribution to my education, more than scientific knowledge, you have passed on values that I will carry with me for the rest of my life.

To the members of the panel for taking part in helping me to complete my degree.

Finally, to everyone who has helped me in any way to complete this stage of my life, my sincere thanks.

SUMMARY

The application of geology to engineering ensures that the geological factors that affect the location, construction and maintenance of civil engineering works are perfectly recognised and applied in order to guarantee their safety. In this context, the aim of this study is to gather data on how geological studies are used in the municipality of Mossoró in Rio Grande do Norte. The procedures adopted were bibliographical and documentary research and the collection of information in the field, carried out with companies that carry out civil construction work in the municipality, from May 2013 to August of the same year. The results obtained indicate that there are studies before, during and after construction, which guarantees the stability, technical and economic viability of civil engineering works in the municipality of Mossoró, thus increasing their safety. Finally, considerations are presented that illustrate the development of civil construction in the city, where housing and road works predominate.

Keywords: Geology - Studies. Construction - Mossoró, RN. Infrastructure works.

SUMMARY

CHAPTER 1

INTRODUCTION

The advance of civil construction in Brazil has led to the execution of major infrastructure projects in the country. However, as the speed of construction increases, the risks become ever more imminent and need to be calculated. This is why it is important to apply geology, as it guarantees the stability, technical and economic viability of civil engineering works, thereby increasing their safety.

Maciel Filho (2008) assures us that Geology Applied to Engineering is of great relevance to Civil Engineering because it is applied to the area in various ways, mainly by providing the geological conditions necessary for the development of Civil Engineering projects so that they can be carried out safely, economically and with technical feasibility.

Major civil engineering works such as road construction, dams and underground works require geological studies of the regions where they will be built, since their permanence over the useful life of the works is related to the geological constraints of the construction site that will need to be met in the executive design of these works, so the geological constraints of the region will provide important parameters for the sizing of these works.

Therefore, the fundamental purpose of engineering geology is to co-operate in the economic and safe design of civil engineering projects and, consequently, to reduce significant material damage and, often, loss of life.

Mossoró, the "capital of the west of Rio Grande do Sul", is undergoing significant economic growth in the construction industry: viaducts, avenues being doubled by the Growth Acceleration Programme (PAC) and huge real estate expansion. With this in mind, the need arose to evaluate the geological study applied in the development of a civil engineering project in Mossoró.

In this context, this work consists of a survey of data on how geological studies are used in the municipality of Mossoró in Rio Grande do Norte.

CHAPTER 2

LITERATURE REVIEW

2.1 IMPORTANCE OF GEOLOGY FOR GEOLOGICAL STUDIES IN CONSTRUCTION

The importance of applying geological knowledge from Applied Geology to Civil Engineering stems from the fact that it will make it possible to reduce costs and delivery times; it will facilitate access to construction materials, favour the use of lower levels of safety and create the possibility of caution and correction of some stability problems that may occur.

This importance also stems from the fact that Geology Applied to Civil Engineering is able to pinpoint geological phenomena that may arise and thus prevent them, thus ensuring the permanence of the works. It is geological knowledge that makes it possible to determine the best and safest sites for the construction of buildings and other civil infrastructures.

In the construction industry, soil is considered to be the cheapest and most abundant material found in nature. It serves as the basis for all civil engineering works, and knowledge of its physical and mechanical properties is of great importance in order to combine safety and economy (MONTEIRO, 2005).

The geological conditions attributed by the regional geology of the site to be built, influence the procedures used for its construction, the materials that are most appropriate, the relief where it will be sized, whether the nature of the soil would be appropriate for the type of construction in question and the probable impacts that the work will have on the region, the environment, society and the economy. The combination of these factors is of great value to the project if it is to be successful.

According to Lollo (2008), specific geological conditions of interest to civil engineers include: soil composition and properties; rock composition and discontinuities; water conditions underground; relief conditions; construction materials present and their properties; soil stability characteristics; and soil excavation dismantling conditions.

The main geological properties of the area need to be clear from the outset of the project, in order to guide the design according to the natural capabilities of the site, enabling the development of a project that is harmonious with the nature of the terrain, economical and safe.

The study of geology is very important for preventing geological accidents. It is by identifying and analysing risk zones that unforeseen events can be avoided. These are then determined through prior work on diagnosing the physical environment, generally known as geotechnical mapping.

According to AMARAL AND CERRI (1998), geological accident prevention measures can be aimed at preventing the occurrence or reducing the magnitude of the geological process(es), at stopping or reducing the resulting social and/or economic consequences, or both at the same time. The authors analyse that, in addition to the possibility of permanently removing residents from areas subject to risk, the prevention of urban geological accidents must take into account the following purposes: eliminating and/or reducing the risks already installed, preventing the installation of new risk areas and living with the current risks.

Geotechnical mapping is a cartographic study of the physical environment, where the search arises for the characterisation of the different aspects: geomorphological, geotechnical, hydrological, environmental and geological, determining the aspects of the materials that exist on the surface and their anthropogenic modifications in the environment. This knowledge, added to the identification of the causes and mechanisms that act on the main geological incidents, provides the necessary elements to guide the process of land use and occupation, improving the occupation of environments according to the definition of the adaptability of the land for different purposes, reducing the risks of these accidents.

In this sense, differentiating areas of geological risk and proposing measures to prevent the corresponding accidents is of paramount importance, with the recommendation of warning sites, their qualification and characteristics, shown on geological risk zoning charts.

2.2   HISTORY OF THESE STUDIES IN THE WORLD AND IN BRAZIL

A major contributor to geological studies worldwide was Karl Terzaghi, who is considered by many to be the father of soil mechanics.

According to Vargas (1983), in 1925 he **published the book "Erbaumechnik auf Bodenphysikalischep Grundlage" and** later published eight articles in the United States under the **general title "Principies of Soil Mechanics".** From 1925 to 1929 he taught at the Massashussets Institute of Technology (MIT), with Arthur Casagrande as his assistant. From 1930 to 1938 Terzaghi taught in Vienna. In 1936, Arthur Casagrande prepared the first International Conference on Soil Mechanics and Foundation Engineering at Harvard. Later, in 1938, Terzaghi returned to the United States to teach Applied Geology at the Harvard Graduate School of Engineering, where he continued teaching until he retired in 1956. In 1947, geological studies in Brazil had the great honour of welcoming Karl Terzaghi to teach six classes in Applied Geology at USP's Polytechnic School.

According to Costa Nunes (1983), a point emphasised in Terzaghi's teachings was the geological aspects, including the details. In the various historical events, the origin of the problems are the geological aspects and at the same time guide the solutions. The geometry of the geotechnical pattern originates from the geological study.

Langer (1990) recognises three phases in the development of engineering geology. The term engineering geology was introduced in 1874 in Austria. In the first phase, research and decision-making were genuinely geological. In Brazil, this phase persisted for years after World War II. The teaching programmes followed this general orientation, that of purely geological positions.

The second phase developed after World War II. Design engineers and consultants analysed the need for more information to establish the correlation between geology and building.

Groups developed that involved geologists, soil and rock mechanists, working together with design engineers and builders, with the aim of safety, profit and technical innovation in construction, both on the surface and underground (Langer, 1990).

Langer (1990) states that in the third phase, environmental concerns were added. For this reason, the International Association of Engineering Geology (IAEG), at its General Assembly in 1980, suggested that all those most qualified in the field of engineering geology, when designing and constructing works, should give their full attention not only to their feasibility and effectiveness, but also, to the same extent, to safeguarding the environment and using it wisely; and in so doing, endeavour to establish quantitative forecasts of the consequences of human activities and natural methods on the geological environment, in terms of space, time, mode and intensity. The consequence of this phase is partly the development of modern technology and industrial technology, and partly society's growing awareness of environmental problems.

According to Maciel Filho (2008), geological studies applied to engineering began in Brazil at the Geological Research Centre for the Inspection of Works against Droughts, in the northeast, where severe droughts were occurring and the need arose to build dams in the region. In addition, the construction of the Ilha Solteira Dam in 1978, located between the cities of Ilha Solteira (SP) and Selvíria (MS), began to play an important role in the process of expanding geology applied to engineering, since it was a project that had major geological problems which forced investments in research in the area and the hiring of professionals specialised in the sector to work for engineering companies.

According to Vargas (in Ruiz, 1987), the first existing information on geology applied to engineering works dates back to 1907, authored by Miguel Arrojado Lisboa, and mentions the expansion of the Noroeste do Brasil railway. Lisboa was also responsible for setting up the Geological Research Centre of the Drought Works Inspectorate in 1909, where he and American geologists carried out numerous geological studies for dam sites in the Northeast.

According to Maciel Filho (2008), as time went by, the Brazilian

Association of Engineering Geology (ABGE) was created, which provided a number of cases and studies in the area that made Civil Engineers realise that it was essential to take into account the knowledge of Geology Applied to Engineering in the sizing of their construction works.

2.3 THE MAIN METHODS OR TECHNIQUES OF GEOLOGICAL STUDIES USED IN CONSTRUCTION.

The main geological methods or techniques in engineering geology can be divided into direct and indirect methods, which deal with the investigation of the subsoil, i.e. ascertaining the layout, nature and thickness of its layers, as well as their properties.

The direct methods aim to carry out excavations with the intention of prospecting the massifs, distinguishing their lithological properties, geotechnical aspects of the materials and geological structures, including Simple Penetration Tests (SPT), boreholes and trenches, auger drilling, percussion drilling, rotary drilling and core drilling.

Indirect methods, on the other hand, are based on field tests that do not alter the physical characteristics of the material being analysed, using morphological, topographical and physical aspects of the terrain. These include geoelectric methods, which are subdivided into electroresistivity, vertical electrical probing, induced polarisation and spontaneous potential; potential methods, which are broken down into magnometry and gravimetry; and seismic methods, which are broken down into reflection and refraction.

According to Maciel Filho (2007), indirect methods gather information by determining certain physical characteristics which, if interpreted correctly, can provide considerable information about the soil and rock bodies. In the case of direct methods, the information is gathered through contact between the researcher and the material to be assessed, which occurs directly, i.e. by taking samples of the material.

2.3.1 DIRECT METHODS

2.3.1.1 Percussion drilling

It is the most common drilling method used in Brazil because it is relatively inexpensive and is used to characterise the land cover of natural soils.

According to Chiossi (1975), percussion drilling only requires a wooden or metal tripod, a small 200-litre water tank and soil cutting tools. It begins with a simple auger to find the water level, and there may be the possibility of the walls of the hole collapsing, especially in sandy soils. The author also states that in soils, sediments or rocks that are not very resistant, this type of borehole is the most commonly used for the construction of buildings, dams, bridges and more.

The Simple Penetration Test (SPT) is a geological investigation method characterised by a standard penetration test that takes place in order to collect a soil sample to be analysed in the laboratory for its composition, soil type and penetration resistance index (MELLO; TEIXEIRA, 1960, apud LIRA, 2013, p.24).

The Simple Penetration Test (SPT) is carried out using a **metal casing pipe, with an internal diameter of 63.5 mm (2.5")**, a tripod equipped with a sheave and cable; steel rods for drilling, with an internal diameter of 25 mm; a motor-pump assembly for circulating water during drilling; iron hammer for driving the drill rods, sampler and casing; auger with a diameter of 100 mm and helical auger with a diameter of 56 to 62 mm and standard sampler with an external diameter of 50.8 mm and an internal diameter of 34.9 mm, with a split body.

As is well known, the SPT is a widely used test, not only in Brazil, but throughout the world as an indispensable tool in preliminary investigations for foundation design.

According to Seed (1985), since it is an instrument that indicates the type of soil, the borehole profile, the water table and soil resistance, and because of its simplicity, strength and quick response time, it seems reasonable to look for elements that will allow its performance to be assessed more credibly using a standardised procedure.

With this objective in mind, research has evolved in recent decades, adding new data on the SPT. As it is a simple test, both from the point of view of execution and interpretation, since its inception there has been an exorbitant spread of geotechnical companies that have begun to use the SPT as a commonly used test in geotechnical circles. Obviously, this popularisation has led to both positive and negative aspects.

The reality is that the lack of standardisation and certification of services related to SPT-type simple penetration drilling brings with it a great potential for financial, social, environmental and energy waste, affecting the sustainability of the societies in which they are implemented.

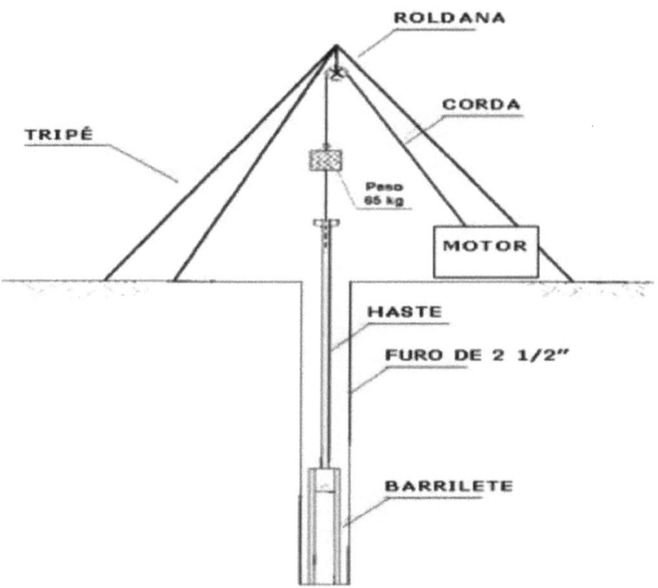

Figure 01: Equipment for percussion test and SPT measurement of subsoil. Available at http://www.forumdaconstrucao.com.br/conteudo.php?a=9&Cod=126

2.3.1.2   Drilling wells and trenches

These are vertical excavations that allow observers access to the interior of the ground so that they can visualise the properties of the mass under study. These types of investigations are useful for collecting deformed and undeformed samples and for carrying out laboratory tests. The depth of the well is restricted by the presence of a water table.

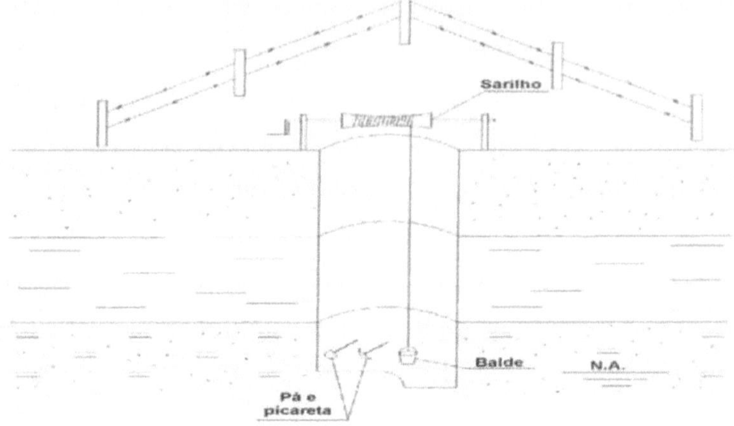

Figure 02: Research well excavated in soil. Source: Oliveira & Brito 1998

2.3.1.3   Auger drilling

This is a manual borehole, usually of short diameter and shallow depth, used to collect samples for soil characterisation tests in the laboratory. It can also be used to determine the **water level. Auger drilling** is done with augers to investigate low and medium resistance soil.

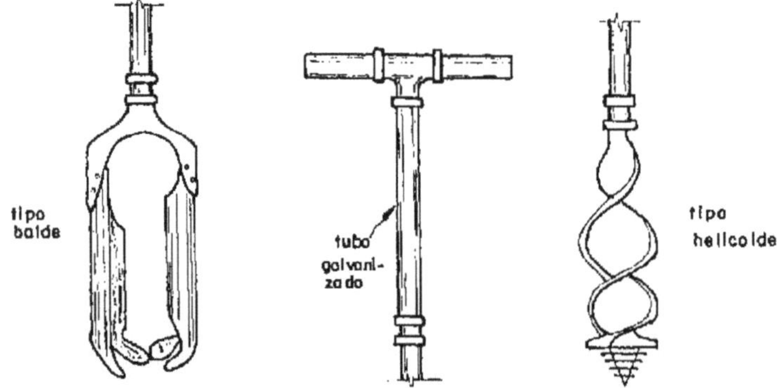

Figure 03: Types of auger. Available at: http://dc182.4shared.com/doc/RkmxuAAI/preview.html

2.3.1.4   Rotary drilling

It is used when you want to know the properties of rocky terrain at great depths, as it can penetrate the rock using a bit capable of drilling through it. The basic equipment for rotary drilling includes a motorised drill rig, water pump, rods, core barrel and core bits.

Rotary drilling is extremely important in soils that have rock extracts, as it is through this drilling that the presence of faults and fractures can be detected, which must be treated to avoid the instability of civil works (MARINHO; 2007, apud LIRA, 2013, p.26).

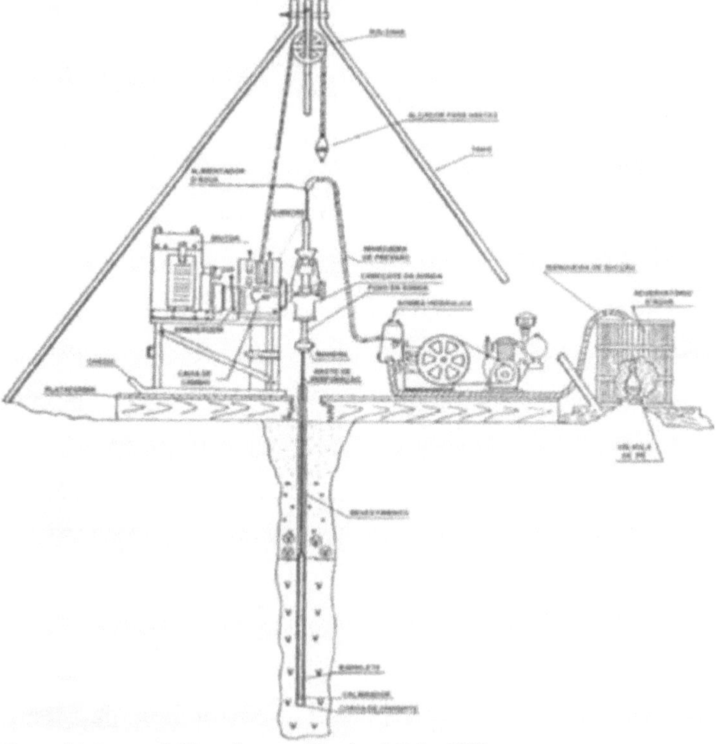

Figure 04: Rotary drilling. Source: Oliveira & Brito 1998.

2.3.1. 5Retail probing

This borehole is carried out with an iron rod, driven manually or with a sledgehammer into submerged unconsolidated sediments. It is used to recognise alluvial deposits, rock surfaces in river beds and to assess sand and gravel deposits for use in construction.

The rod generally penetrates up to 2 metres into the unconsolidated sandy alluvium and the material passed through can be identified by the sound and vibration reaction of the process. In clay, penetration is soft, in sand rough and in sand and gravel deposits sparse blockages are observed

when driving the rod. Consolidated gravel banks are not penetrated by the rod. On a rocky surface, the impact is hard and slips (SOUZA et al., 1998).

2.3.2 INDIRECT METHODS

2.3.2.1 Geoelectric methods

They are those resulting from the physical parameters obtained using appropriate equipment. They aim to characterise and identify the different geological materials in order to achieve the research objectives. They are subdivided into electro-resistivity, vertical electrical sounding, induced polarisation, spontaneous potential and ground penetrating radar.

2.3.2.2 Electroresistance

It is a geophysical method whose principle is based on determining the electrical resistivity of materials which, together with the dielectric constant and magnetic permeability, fundamentally express the electromagnetic properties of soils and rocks. This method makes it difficult for electric currents to propagate in any medium. The data is shown in different ways, such as sections, apparent resistivity isovalue plots and profiles.

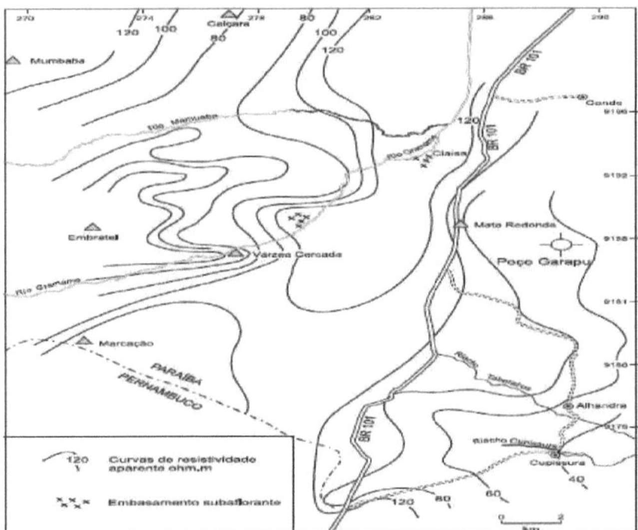

Figure 05: Electrorresistivity map (reconnaissance scale) of the westernmost sedimentary part of the Pernambuco - Paraíba Basin. Available at:
http://ppegeo.igc.usp.br/scielo.php?pid=S1519-874X2009000200004&script=sci arttext

2.3.2.3 Vertical electrical sounding

It consists of a series of measurements of a geoelectric parameter, carried out from the surface of the ground, maintaining an increasing separation between the current emitting and potential receiving electrodes. This method can be used in the construction of large civil works (dams, tunnels and harbours), contaminated areas and landfill sites.

Vertical electrical sounding (VES) is an investigation technique that refers to the geophysical method of electro-resistivity, which is characterised by vertically quantifying, based on an artificially created electrical field, the physical property (electrical resistivity) of the geological environment (lithotypes), relative to its capacity to conduct or resist the propagation of electrical currents. Its application is preferably adopted in flat and gently undulating areas, where the geological layers are believed to be arranged in a horizontal to sub-horizontal manner (SOUZA, 2002).

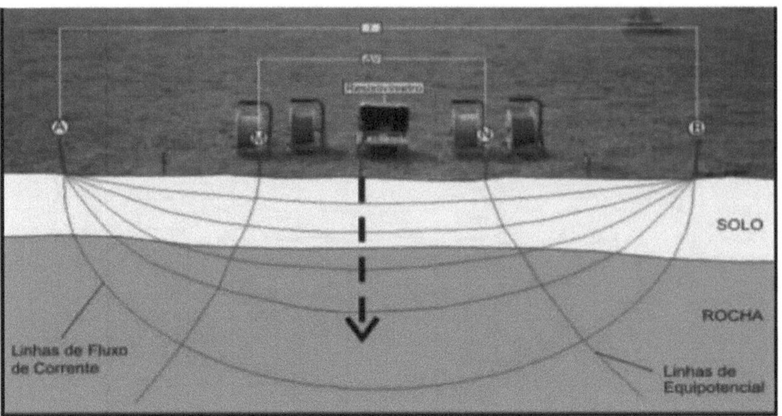

Figure 06: Vertical Electrical Sounding Technique - SEV, Schlumberger Arrangement. Available at: http://www.tecgeofisica.com.br/conteudo/diagnostico.htm

2.3.2. 4Spontaneous potential   and induced polarisation

These are methods that investigate polarisation, which can be spontaneous or induced. They are used on the ground, but the potential difference is measured after the current has stopped or by varying the frequency. They are also used to determine the direction and direction of flow of underground fluids.

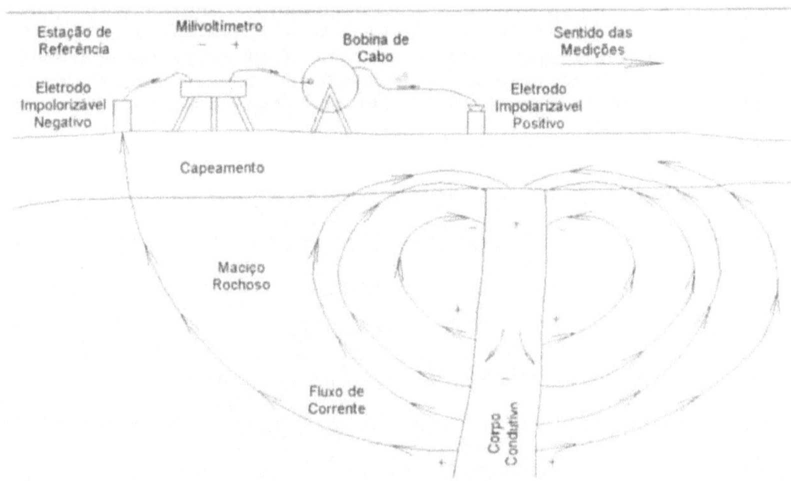

Figure 07: Spontaneous potential. Source: Oliveira & Brito 1998.

2.3.2.5 Potential methods

The two potential methods are divided into magnetometry, which aims to measure small variations in the intensity of the Earth's magnetic field as a result of the irregular distribution of magnetised rocks on the surface, and gravimetry, which is based on determining the gravitational attraction at a point on the Earth's surface, using equipment called gravimeters.

The magnetometric method has many similarities with the gravimetric method. Both relate to natural force fields governed by fundamentally similar laws; they also depend on intrinsic rock properties: density in gravimetry and magnetic susceptibility in magnetometry. (CHIOSSI, 1975, P. 155)

2.3.2. 6Seismic methods

Seismic methods have great applicability in engineering geology, which in turn studies the depth distribution of the acoustic wave propagation velocity parameter, which is closely linked to the physical properties of the geological environment, such as: density, elastic constants, porosity, mineralogical and chemical composition, water content and confinement stress.

The two best-known seismic methods are reflection, which refers to the ability of waves to penetrate a reflector in the subsurface with an inclination less than the critical angle; and refraction, which is based on recording the refracted seismic waves on the surface of lithological contacts or other surfaces.

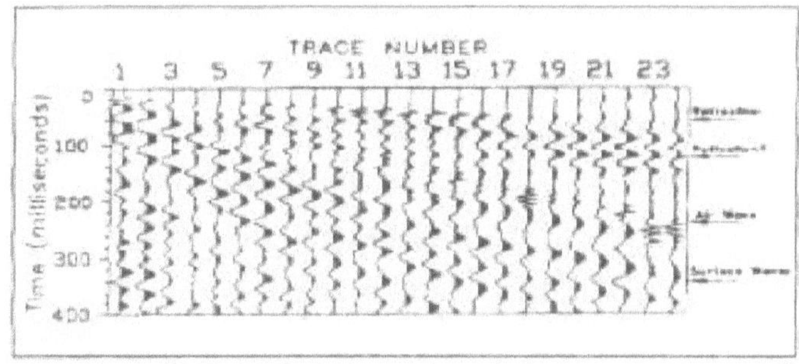

Figure 08:   Seismic reflection record. Available at:
http://geo.web.ua.pt/index.php?option=com content&view=article&id=165%3ASeismic
methods&catid=36%3Ageophysical methods&Itemid=60&limitstart=2

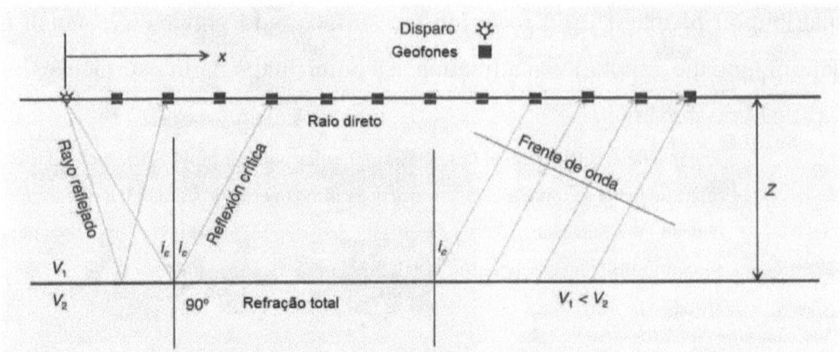

Figure 09:   Refraction seismic diagram. Available at:
http://geo.web.ua.pt/index.php?option=com content&view=article&id=165%3Ametodos-sismicos&catid=36%3Ametodos-geofisicos&Itemid=60&limitstart=1

2.4 MAPPING   THE PHYSICAL AND GEOLOGICAL CONSTRAINTS OF THE MOSSORÓ REGION.

According to IDEMA (1999), the municipality of Mossoró-RN is located in the range of the Potiguar Basin and the Barreiras Group, where light grey and yellowish bioclastic calcarenites and calcilutites predominate, with evaporite levels at the base, deposited on a wide tidal plain and on a shallow platform of the Jandaíra Formation (Apodi Group) of Cretaceous age, 80 million years old.

Belonging to the Apodi Group, the Jandaíra Formation is a carbonate sequence conformably overlying the Açu Sandstone, which is partially covered in the emerged portion of the basin by rocks of the Barreiras Formation and the Açu Formation (Apodi Group), formed by dense clusters of sandstones ranging from very fine to conglomeratic, interspersed with shales, siltstones and claystones (TEÓDULO, 2004).

In the northern part, unconsolidated sediments of the Barreiras Group of Tertiary age predominate, demonstrated by sandstones, unconsolidated sandstones, conglomerates and claystones that develop sandy soils.

yellow and reddish coloured massifs. In the valleys of the Apodi-Mossoró riverbeds there are alluvial deposits made up of sands and gravels, with pelitic intercalations, associated with the current fluvial systems, forming a fluvial plain, a flat area resulting from fluvial accumulation subject to periodic flooding (IDEMA, 1999).

According to IDEMA (1999) among the soils, and according to the new EMBRAPA soil category, Cambissolos dominate, a highly fertile soil with a clayey texture, generally shallow, moderately drained and with flat relief, predominantly hyperxerophilous caatinga; Chernosols, also of high fertility, clay texture, moderate, imperfectly drained and flat relief, and Latosols, occurring on relief with a small slope, where the vegetation is composed of hyperxerophilic caatinga with a predominance of trees, has medium to high fertility and is extremely drained.

Although the seat of the municipality is situated at an altitude of 16 metres, the relief has an average altitude of 100 metres, where the Chapada do Apodi is present, a flattened surface with flat and gently undulating relief (RADAMBRASIL apud TEÓDULO, 2004), formed by sedimentary soils cut by the Apodi-Mossoró and Piranhas-Assú rivers; the River Plains, low, flat terrain located on the banks of the rivers; the Sublittoral Depression, low-lying terrain located between two higher forms of relief and the Depression.

The municipality of Mossoró is mainly crossed by the Apodi-Mossoró and Rio do Carmo rivers, whose valleys are developed transversely to the main regional structural lines, and where they are laid over the sediments of the Açu Formation and the Jandaíra Formation as they approach the Areia Branca coast (TEÓDULO, 2004).

The Apodi-Mossoró River runs through the city in a SW/NE direction, having a tortuous aspect with different lagoons near its banks, narrowing towards the city centre. According to data from IDEMA (1999), the Mossoró region is made up of the main aquifers:

The Jandaíra Aquifer, made up predominantly of limestone, has generally brackish water and a chemical composition favourable to small-scale irrigation, with wells with an average depth of around 8m.

The Assú Aquifer occurs at the edge of the Potiguar Basin and is free in its outcrop. It has an average thickness of 150 metres in the outcrop area.

Among surface resources, the municipality also has the following streams:

Bonsucesso, Cabelo Negro, São Raimundo, Pai Antônio (IDEMA, 1999).

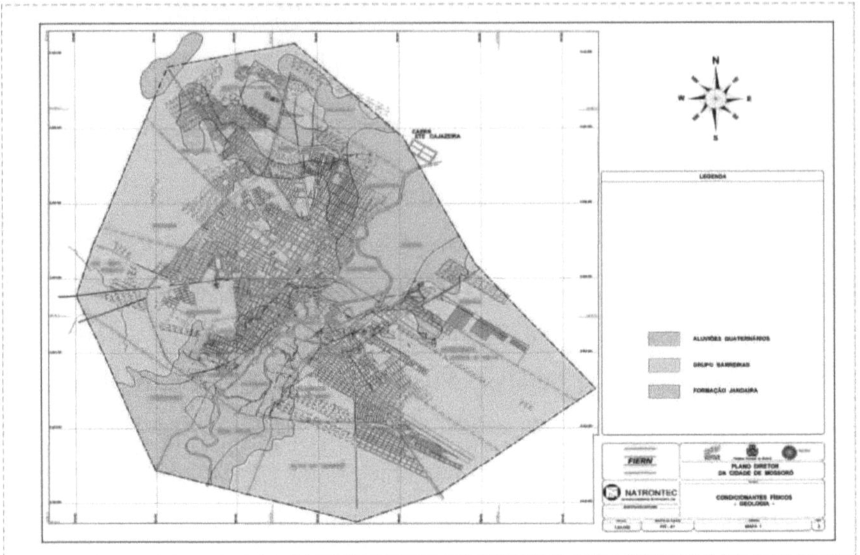

Figure 10: Physical constraints - geology. Source: Secretaria Municipal de Desenvolvimento Urbano: Relatório de condicionantes físicos e naturais de Mossoró, 1997.

2.5   CIVIL CONSTRUCTION IN THE MUNICIPALITY OF MOSSORÓ-RN.

Situated between Natal and Fortaleza, to which it is linked by the BR-304 motorway, Mossoró is one of the main cities in the northeastern interior according to the location of (figure 11) below, and is currently experiencing major economic and infrastructure growth, rated as one of Brazil's most attractive medium-sized cities for investment. The municipality is the country's largest producer of onshore oil, as well as sea salt.

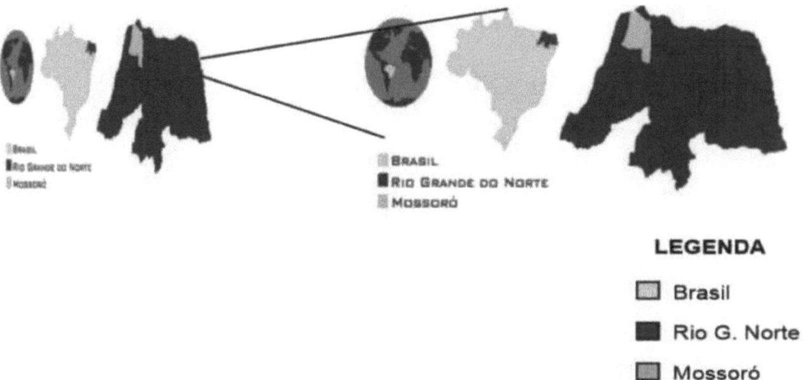

Figure 11: Map showing the location of the study area. Source: Revista Verde de Agroecologia e Desenvolvimento Sustentável Grupo Verde de Agricultura Alternativa (GVAA), 2007.

In the process of urbanisation, the great challenge for cities is urban growth and development that generates wealth, quality of life and environmental quality for their present and future inhabitants (ARAÚJO and CARAM, 2006).

Nowadays, the city of Mossoró is experiencing great economic and infrastructural growth, analysed among medium-sized cities as a good alternative for investment. As far as construction is concerned, a few years ago there weren't many high-rise buildings in the city. Nowadays you can see large buildings, because the city stands out from other cities in the interior because it has excellent job opportunities, several higher education institutions, technical schools and a well-established economy. The sum of these attractions has caused urbanisation to accelerate and consequently the city's real estate sector to heat up.

Figure 12: Vertical construction of a building in the centre of Mossoró-RN, 2013. Source: Mibson Michel

Civil construction in Mossoró-RN is undergoing diversified growth in the real estate and infrastructure sectors, such as the vertical construction of many buildings in the various neighbourhoods that exist in the city, according to (Figure 12) above. Another form of construction that is widespread in the Mossoró area is its infrastructure work related to traffic engineering, which would be the construction of the city's road complex with a total of 5 viaducts, as shown in (Figure 13) below, and the restructuring and duplication of 17 kilometres from the city on the BR-304/RN up to the entrance to the municipality of Tibau-RN.

Figure 13: Construction of a viaduct around the municipality of Mossoró-RN, 2013. Source: Mibson Michel.

CHAPTER 3

METHODOLOGY

3.1 RESEARCH PROCEDURE

The procedures adopted were bibliographical and documentary research and data collection in the field, with companies that carry out construction work in the municipality of Mossoró-RN, from May 2013 to August of the same year.

3.2 BIBLIOGRAPHICAL RESEARCH

Books, reports and articles on the following subjects were used: geology applied to engineering, geological studies in civil construction, geological research, geotechnics, geological accidents. We also consulted monographs from the Federal University's library.

Rural do Semi-Arido (UFERSA), Mossoró campus. In addition, visits were made to public institutions, such as the Mossoró City Hall (Municipal Secretariat for Urban Development) and the Vingt-un Rosado Foundation.

3.3 DOCUMENTARY RESEARCH

The methodology adopted in this study was to analyse the registration of companies with the Mossoró Construction Industry Union (SINDUSCON) in order to obtain information on the number of companies carrying out construction work in and around the municipality of Mossoró-RN. Therefore, the methodology of this research was firstly based on theoretical foundations, consisting of a survey of bibliographic analysis, and then field data collection and photographic recording of vertical building constructions in general, the Public Prosecutor's Office of Rio Grande do Norte in

Mossoró (MP-RN) and works under the Growth Acceleration Programme (PAC) around Mossoró, such as its viaducts and road duplications.

3.4   POPULATION AND SAMPLE

The questionnaire (Appendix A) was designed to be applied to construction companies in the municipality of Mossoró. The companies chosen to administer the questionnaires were randomised. At the same time as the questionnaires were administered, photographs were taken of the application of engineering and geology to buildings related to the theme.

In this research, the population considered is 44 construction companies in the municipality of Mossoró - RN, according to the Mossoró Construction Industry Union (SINDUSCON), through the registration of these companies with the union.

The sample size was defined as indicated by (RICHARDSON, 1999), using the following formula:

$$n = \sigma^2 . \text{p. q.} [E^{\top}(2)\ (N-1) + \sigma^2 . \text{p. q}]^{\top}(-1)$$ where:

n: Sample size;

N: Population size;

σ^2 : Confidence level, in number of deviations;

p: Proportion of the characteristics surveyed in the universe, given as a percentage;

q: Proportion of the universe that does not have the characteristic being researched (q =1- p), transformed into a percentage;

E: Estimation error allowed.

With an estimated standard error of around 10 per cent, a population universe (N) of 43 companies and a confidence level **(a^2) of 90 per cent, the data is as follows:**

$$N = 1^2 . 50. 50. [10^2 . (44\text{-}1) + 1^2 . 50.50]^{-1}$$
$$N = 2500. [100. (43) + 1. 2500]^{-1}$$

$$N = 2500 . [4300 + 2500]^{-1}$$

$$N = 2500 . [6800]^{-1}$$

$$N = 2500 / 6800$$

$$N = 0, 3676$$

$$N = 0,3676\%$$

$$N = 0.3676\% \times 44$$

$$N = 16,1744$$

$$N = 16 \text{ Companies}$$

According to Richardson's formula, the number of questionnaires to be administered to companies carrying out construction work in Mossoró was 16.

CHAPTER 4

RESULTS AND DISCUSSION

Through the application of questionnaires to construction companies in Mossoró - the universe of the research - it was possible to make clear aspects of geological studies, the type of construction carried out in the municipality and analyses or investigative methods of the subsoil that are directly related to engineering and geology and their direct effects on the choice of the type of method to be used in a construction project. To this end, information was gathered on companies, geological studies carried out, the cost of such studies before, during and after a construction project and the difficulty of using such geological studies in the field of engineering, among other things.

4.1 COMPANY PROFILE AND IDENTIFICATION

As shown in Graph 01, the 16 companies interviewed in Mossoró-RN were aimed at employees and people who held technical positions and were knowledgeable about construction and civil engineering techniques, such as engineers, technicians and those responsible for the work. The companies interviewed in the survey are diversified into large, medium and small, and all play a fundamental role in construction in Mossoró today.

Graph 01: Profile and identification of the companies interviewed in Mossoró-RN, 2013.

Graph 02 shows the predominant type of construction in the city of Mossoró. It was found that half of the construction is housing, or 50%, while water works account for only 6%. Road, commercial and other works account for 13%, 12% and 19% respectively. Other construction includes oil and water tanks, both of which are mineral resources used extensively in the city's growth.

The increase in housing construction is due to the rise in the purchasing power of Brazilians and this can be seen throughout the country in the incentives for buying, building and renovating such properties.

According to CREA-AM (2013), housing credit will be the largest type of credit in 2013 because Brazil still has a small ratio of housing credit to GDP at 6.2%, unlike developed countries where this percentage exceeds 50% and the level of construction in the building sector is in high demand.

Mossoró and the country as a whole are demonstrating that the growing and continuous development of residential buildings still has a lot of scope for development.

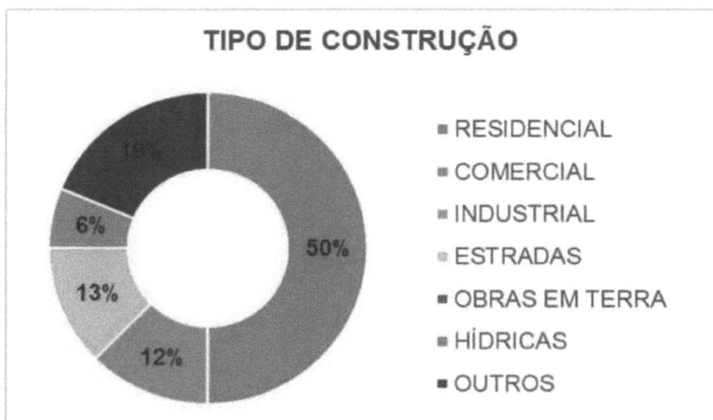

Graph 02: Representation of the predominant type of construction in Mossoró-RN, 2013.

4.2 THE RELATIONSHIP   BETWEEN GEOLOGICAL STUDIES AND CIVIL CONSTRUCTION

With regard to the method of drilling and subsoil investigation (Graph 03), the vast majority (94%) of those interviewed use the direct method as the most common in the municipality of Mossoró, while the remaining 6% use the indirect method, used by only one company out of the 16 in question, which is Vertical Electrical Boring. Of the methods discussed, the Simple Penetration Test (SPT) stands out with enormous adherence because, according to those surveyed, it is a very common method among companies and because it is affordable, simple and indicative of the characteristics of the site.

The SPT, also known as percussion or simple reconnaissance drilling, is undoubtedly the most widely used test in most countries, including Brazil. It should be noted that in recent decades there has been an inclination to switch to the SPT-T, with the same relative cost and greater complexity (HACHICH, 1996).

Compared to Seed (1985), who reports that the SPT indicates the type of soil, the borehole profile, the water table and the resistance that the soil offers to a particular work that overloads it, and because it is quite simple even in its accelerated response time and standardisation in carrying out the method.

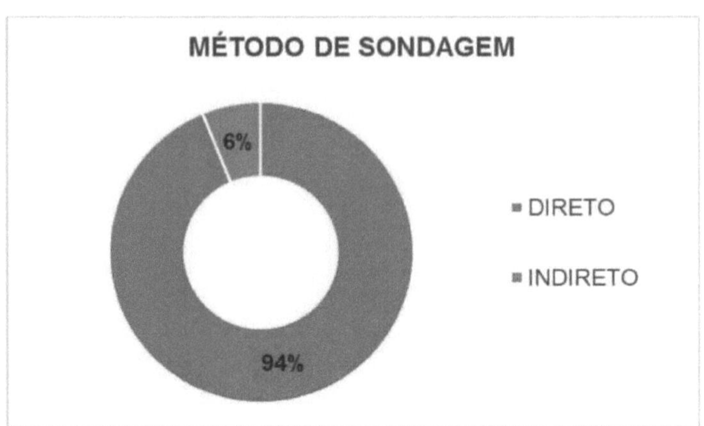

Graph 03: Boring method used on construction sites in Mossoró-RN, 2013.

When asked "How important is a geological study before, during and after construction?", the answers ranged from preventing problems with clay soils settling, excess water at shallow depths and, in some cases, low water absorption after the soil has been moistened, highlighting the concern with predicting geological accidents.

Another important survey was on the type of soil that predominates at the study site. According to IDEMA (1999), according to the EMBRAPA classification, the soil is dominantly quite fertile, with a clay texture, low drainage and a reddish-yellow colour characteristic of the region's claystones. Confirming what was seen in the field through the data obtained (Graph 04), 56% of the soils are clayey, 25% sandy, 13% silty and 6% limestone.

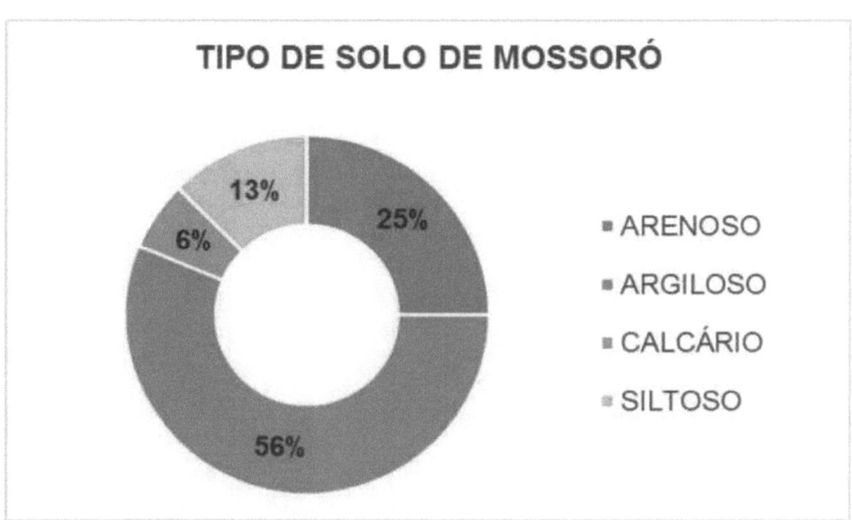

Graph 04: Type of soil found in the municipality of Mossoró-RN, 2013.

As for the state of construction in Mossoró (Graph 05), 38 per cent of the growth was considerably good, 31 per cent excellent, thus revealing that the city's construction sector is, in the opinion of the city's builders, a good option for work, with 25 per cent reasonable and 6 per cent poor. Compared to the results of Freire's research (2011), in the last ten years Mossoró's property business has grown by an incredible 450%, compared to the national average of 43.2%. Despite this, the city has great market potential for the next ten years. The case and the optimistic forecast are made by the Mossoró Construction Union (Sinduscon). Mossoró's performance in comparative terms puts the municipality in an extraordinarily high position when based, for example, on the development of verticalisation in the major Brazilian regions over the last ten years separately, as shown in (figure 14) below.

Figure 14: Construction process in the municipality of Mossoró. Available at:
http://www.skyscrapercity.com/showthread.php?t=1591321&page=3

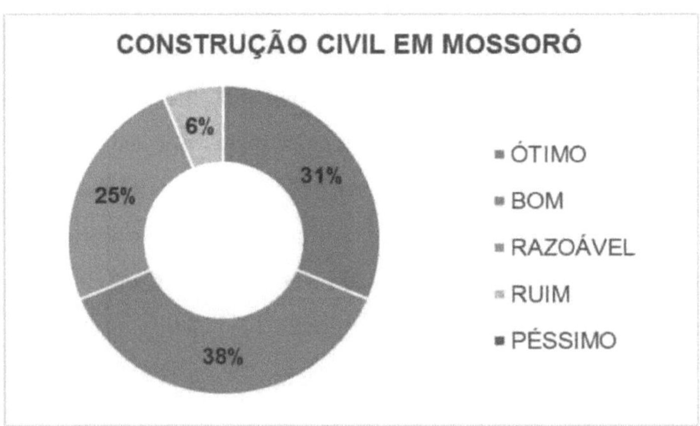

Graph 05: Growth of civil construction in the municipality of Mossoró-RN, 2013.

When **asked about the "Main difficulty in using** geological study methods", the interviewees reported that there was a possible significant increase in the cost of the work in their building budgets and a difficulty in finding companies specialised in the field and with the right machinery for the various civil engineering projects in the municipality.

The difficulty in using geological study methods becomes **more pronounced with the survey on "Is the geological study carried out by the** company itself or by an outsourced company?" which explains that all the companies unanimously answered that this study is carried out by outsourced companies and that in Mossoró there is no such company in the area, leaving companies from Fortaleza-CE and Natal- RN to carry out the service because they are the **closest** capitals **to the "Land of Sun, Salt and Oil".**

The result of outsourcing the geological study to outside companies would be to increase the quality of the service and reduce costs in the long term with repair work and minimise future repairs with geological accidents due to poor execution in the initial building process. As for the difficulty of finding these companies in Mossoró, it would be because it is a complex and innovative field with interaction between engineering, geology and computing, so that through advanced equipment they can have greater precision in the execution of the project.

CHAPTER 5

FINAL CONSIDERATIONS

The conclusion of this research is that Civil Engineering needs to go hand in hand with Geology Applied to Engineering, because it is through the geological knowledge of the region provided by Geology Applied to Engineering that the engineer will achieve technical and economic viability in his works, maintaining their stability throughout their useful life.

It was also possible to see that in the municipality of Mossoró there is only one geological study before and during construction, due to the concern with the sizing of the foundations of the works, and that carrying out this analysis is not so acceptable due to the lack of knowledge of the study method applied to geology and the fact that it makes the construction budget more expensive.

Finally, it was concluded that **the "capital of western Potiguar"** is undergoing major and significant economic growth in the civil construction sector, where housing and roadworks in the city's surroundings predominate. However, it is necessary for companies to take into account all the geological constraints of the region, so that in the future there are no problems in their work that cause instability.

CHAPTER 6

REFERENCES

ARAÚJO, Bianca Carla Dantas de; CARAM, Rosana. **Environmental Analysis: Urban Bioclimatic Study in a Historic Centre.** Ambiente & Sociedade - Vol. IX n°. 1 jan./jun. 2006.

CREA-AM, **Housing credit will be the largest credit modality in 2013.** Available at: <http://www.crea-am.org.br/src/site/noticia.php?id=2702>. Accessed on 06 September 2013.

FREIRE, Paulo Sérgio. **Verticalisation grows above average.** Available at: <http://tribunadonorte.com.br/news.php?not id=189930>. Accessed on 27 August 2013.

HACHICH, W.R.; TAROZZO, H. (1996). **Foundations: theory and practice.** São Paulo: PINI Publishing House.

Institute for Economic Development and the Environment-IDEMA, 1999. Available at: <http://www.idema.rn.gov.br/perfildoseumunicipio>. Accessed: 23/04/2007.

Langer M. 1990. **The géologie de l'ingénieur aujourd'hui: exigences et réalités. Bulletin de l'Association Internationale de Géologie de L'Ingénieur,** Paris, n.42, p. 123-126,

MELLO, Victor F. B.; TEIXEIRA, Alberto H. **Mecânica dos Solos, Fundações e Obras de Terra.** São Paulo: São Carlos School of Engineering, 1960.

Oliveira, A.M.S. & Brito, S.N.A. 1998. **Engineering Geology.** São Paulo, Brazilian Association of Engineering Geology-ABGE, CNPq/FAPESP, 586p.

MOSSORÓ MUNICIPAL GOVERNMENT-PMM. **Municipal Department of Urban Development:** Report on the physical and natural conditions of Mossoró. 1997, p.23 (Electronic Document).

RICHARDSON, Roberto Jarry. **Social Research: Methods and Techniques.** 3 Ed. São Paulo: ATLAS, 336p, 1999.

SEED, H.B.; TOKIMATSU, K.; HADER, L.F.; CHUNG, R.M. **Influence of SPT Procedures in Soil Liquefaction Resultance Evaluations.** Journal of Geotechnical Engineering. ASCE. Vol. 111, No. 12. 1985.

TEÓDULO, José Macio Ramalho. **Use of geoprocessing and remote sensing techniques in the collection and integration of data necessary for the environmental management of the oil and gas extraction fields of canto do Amaro and Alto da Pedra in the municipality of Mossoró-RN.** Master's dissertation, Natal/RN, 2004.

Vargas M. 1983. **Karl Terzaghi and Brazil. In: ABMS Karl Terzaghi: aspects of his life and work.** Rio de Janeiro: ABMS. P. I-1 to I-8.

CHAPTER 7

APPENDIX A

QUESTIONNAIRE APPLIED TO CONSTRUCTION COMPANIES IN MOSSORÓ-RN.

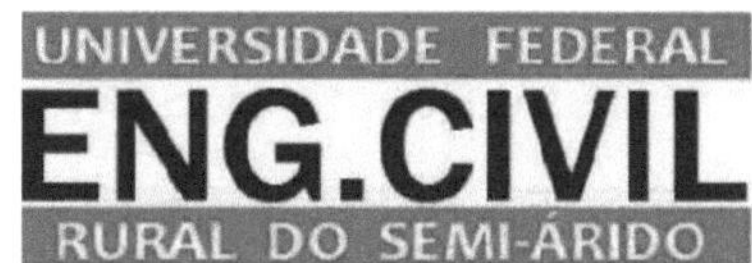

FEDERAL RURAL UNIVERSITY OF THE SEMI-ARID - UFERSA

DEPARTMENT OF EXTERNAL AND NATURAL SCIENCES

BACHELOR'S DEGREE COURSE IN SCIENCE AND TECHNOLOGY

Please answer the questions below. Please note that the purpose is strictly academic and under no circumstances will the information provided be used for any other purpose. We would like to point out that it will form part of the Final Paper for the Bachelor's Degree in Civil Engineering at UFERSA/Mossoró-RN. Thank you in advance for your valuable co-operation.

Student: Mibson Michel Santiago Ramos

Supervisor: Prof. Dr. sc. Marcelo Tavares Gurgel (UFERSA-DCAT).

Date: ___/___/___

COMPANY PROFILE AND IDENTIFICATION

1. COMPANY NAME: ____

2. PROFESSION: ____

3. POSITION HELD IN THE COMPANY: _______________________________

LIST OF GEOLOGICAL STUDIES IN CONSTRUCTION

4. WHAT TYPE OF CONSTRUCTION IS PREDOMINATING IN MOSSORÓ

   ( ) RESIDENTIAL ( ) COMMERCIAL ( ) INDUSTRIAL ( ) ROADS ( )

WORKS

ON LAND ( ) WATER ( ) OTHER: _______________________________________

5. WILL GEOLOGICAL STUDIES AND MONITORING BE CARRIED OUT BEFORE, DURING AND AFTER CONSTRUCTION? ________________

6. THE MAIN DIFFICULTY IN USING GEOLOGICAL STUDY METHODS?

7. WHAT IS THE IMPORTANCE OF A GEOLOGICAL STUDY PRIOR TO A PROJECT?

CONSTRUCTION? _______________________________________

8. GROWTH IN CIVIL CONSTRUCTION IN MOSSORÓ-RN IS FOUND:

( ) GREAT ( ) GOOD ( ) REASONABLE ( ) BAD ( ) TERRIBLE

9. WHAT TYPE OF PROBLEM DOES THE SOIL POSE IN MOSSORÓ? ___

10. WHAT TYPE OF SUBSOIL INVESTIGATION METHOD IS USED? ( ) DIRECT ( ) INDIRECT, AND WHICH ONE? _______________________________

11. IS THE GEOLOGICAL STUDY CARRIED OUT BY THE COMPANY ITSELF OR BY A THIRD PARTY?

12. TYPE OF SOIL FOUND IN THE MUNICIPALITY OF MOSSORÓ:

( ) SANDY ( ) CLAY ( ) SILTY ( ) LIMESTONE

I want morebooks!

Buy your books fast and straightforward online - at one of world's fastest growing online book stores! Environmentally sound due to Print-on-Demand technologies.

Buy your books online at
www.morebooks.shop

Kaufen Sie Ihre Bücher schnell und unkompliziert online – auf einer der am schnellsten wachsenden Buchhandelsplattformen weltweit! Dank Print-On-Demand umwelt- und ressourcenschonend produzi ert.

Bücher schneller online kaufen
www.morebooks.shop